TEACHERS' NOTES

Horrible Science Teachers' Resources: Electricity is inspired by the Horrible Science book *Shocking Electricity*. Each photocopiable takes a weird and wonderful excerpt from the original, as well as from *Suffering Scientists*, *Explosive Experiments*, *Frightening Light* and *The Awfully Big Quiz Book*, and expands on it to create a class-based teaching activity, fulfilling both National Curriculum and QCA objectives. The activities can be used individually or in a series as part of your scheme of work.

With an emphasis on research, experimentation and interpreting results, the activities will appeal to anyone even remotely curious about the Horrible world around us!

PART 1:
ELECTRICITY AROUND US

Page 11: Our electricity
Learning objective
Everyday appliances use electricity.

Start this session by playing a cassette or CD, asking the children 'How is this powered?' Encourage the children to think of other electrically powered items and compile a class list. Use photocopiable page 11 to focus your class on the role electricity plays in our everyday life – and to imagine what life would be like without it! When drawing electrical items encourage the children to include plugs where appropriate. **SAFETY NOTE!** During the discussion, make your class aware that playing with electricity, plugs and sockets is very dangerous and can be fatal!

Page 12: Mains and Battery
Learning objective
That everyday devices are connected to the mains and that they must be used safely.
That some devices use batteries which supply electricity; these can be handled safely.

Use the Horra Towers letter (photocopiable page 12) to focus your class on different uses for electricity. Use a classroom example of a mains-powered item (e.g. computer) and contrast this with a battery-powered item (e.g. digital watch). Ask the children to consider why they use different power sources (practicality, amount of power required by item). Discuss with your class the reasons why some items use more electricity than others and the associated safety implications.
When compiling the lists for Horra Towers,

encourage the children to add other items and to classify them as mains or battery powered.

Pages 13 & 14: Power cut and The power cut continues!
Learning objective
Thinking about what might happen.
That an electrical device will not work if there is a break in the circuit.

Make sure that your classroom lights, CD player, and computer are all switched on when the children enter the room. Switch each item off one by one, asking the children to describe what you are doing. Tell them that doing without electricity is how life would be like in a power cut. Divide your class into small discussion groups to read photocopiable page 13 and encourage them to make notes about how a power cut would affect their lives. You can use these notes as the basis for a class discussion.
Ask the children to continue the cartoon on photocopiable page 14 to include the next day, breakfast, going to school etc. The group notes from earlier can also be useful for adding footnotes to the new cartoon, as in the original.

Page 15: Safety First!
Learning objective
Use a wide range of methods to communicate in an appropriate and systematic manner.
That everyday appliances are connected to mains and that they must be used safely.

Look at photocopiable page 15, explaining to your class that although the adverts are jokey, the inventions were genuine! Talk about why these inventions might have been dangerous and encourage the children to take on the role of safety officer, identifying danger points (too many wires from a socket, water and electricity etc). Move the discussion on to electrical safety in the classroom and also at home, focusing on how we can all use electricity responsibly.

Ensure that all the children are fully aware of the dangers of mains electricity before introducing the safety poster activity. Other posters (on any subject) will be useful in identifying key features (slogan, eye-catching graphic, bullet points, text etc).

PART 2:
MOVEMENT

Page 16: Spot the electric motor!
Learning objective
Ask questions that can be investigated scientifically and decide how to find answers.
That everyday appliances use electricity; these include things that light up, heat up, produce sounds and move.

Start by asking the children to describe the sound a fridge makes. Explain that this is the sound of an electrical motor and ask 'What is it helping the fridge to do?' Introduce photocopiable page 16 and ask the children to work individually or in pairs to classify each item and to add their own examples. At the end of the session encourage the children to share their suggestions and compile a class list.
Answer: All of them!

Page 17: How to make an electric motor!
Learning objective
Make systematic observations and measurements.

Start by identifying the main features of a written experiment. Many children will have followed simple cookery recipes so draw links between the two. Add scientific names to the headings on photocopiable page 17: 'What you need is', 'What you do is' and 'What do you notice' ('Equipment', 'Method' and 'Results', respectively). Explain to the children that this is how scientists always record their experiments. Recap any work you have done about the electric motors in items we use every day. Divide your class into groups or pairs and supervise the main activity

as appropriate. Afterwards, encourage the children to record their experiment and observations with diagrams to help as necessary.
Answer: b) the compass needle turns round and round and the needle twists around. Either way the magnetic field produced by the wire moves with the wire. This keeps pushing away and then attracting the magnetic needle – just like a real electric motor!

Page 18: How to make an electro-magnet!
Learning objective
Use simple equipment and materials appropriately. To make and record observations.

If your class has experienced the previous activity, start by recapping the important role that magnets play in electricity. Explain that this is because of the effect they have on atoms. Electrons make up the electrical current and electrons move them along (a really good explanation of this is on pages 15 and 16 of *Shocking Electricity* – 'Meet the Atom Family'). Encourage the children to work in pairs to make their electro-magnet, recording their thoughts as they progress. Ask for volunteers to talk about this discovery process afterwards.

PART 3:
PARTS OF A CIRCUIT

Page 19: The Battery
Learning objective
To make connections in circuits to the positive and negative poles of a battery.
That a circuit needs a power source.

Start by drawing a selection of electrical items, making sure to include a torch, on the board. Ask the children to identify the items that may be battery powered. If you have a collection of different batteries (digital watch, radio, variety of sizes and strengths) pass them around, asking the children to describe them.

Use photocopiable page 19 so that your class can look inside a battery and see how it works, explaining any new vocabulary. A word bank for new words would be useful at this point. Split the children into pairs and give each a torch. Ask the class to dismantle the torches to see how they fit together, recording their observations and focusing on the + and − signs on the battery cells. Identify the different components on the torch and challenge the children to reassemble them so that the bulb lights up again.

SAFETY NOTE! Make sure all batteries are new and do not leak. Explain that it is not possible to cut open a battery cell as the acids inside are dangerous and will burn the skin. Early batteries were made of cardboard and regularly leaked!

Answer c) Remember that negative electrons flow towards positive atoms – so for electricity to flow and your torch to work you have to put the negative and positive ends together.

Page 20: The Light Bulb
Learning objective
To construct simple circuits.

Encourage your class to name as many different types of bulb as they can (camera flash, bedside lamp, fridge lights, class lights etc). Use photocopiable page 20 as a starting point to identify important parts of the bulb, focusing on the filament. A domestic light bulb would be useful at this point, as the children will be able to see whether the bulb is likely to work or not. Test out the bulb to see if they were right.

Ask the children to draw a light bulb on the photocopiable sheet from a real example (a torch bulb will do) and help them attach the labels. Use the 'Bet you never knew!' information as a starting point to research other, more obscure uses for light bulbs.

Page 21: Selling Light Bulbs!
Learning objective
To use a wide range of methods to communicate in an appropriate and systematic manner.

Explain to your class that people were wary when light bulbs were first invented, as early attempts were expensive and sometimes shattered! Wealthy people had already fitted gas pipes for light and did not want to start all over again.

Put the children in charge of persuading the public to go electric!

Read the newspaper report on photocopiable page 21 together and use the headings below it to plan a promotional pamphlet. Pamphlets on any subject can be used as examples of this style of writing. Encourage the children to present their pamphlets when completed, adapting some into TV-style commercials. You can use the pamphlets themselves as a class display.

Page 22: Conductors
Learning objective
That some materials are better conductors of electricity than others.

Demonstrate to your class how they can make a simple circuit with a bulb, battery, wires and metal clips (there are other activities that focus on different types of circuit in Part 4: Making Circuits). Explain that the clips are metal and ask the children why they think wooden clothes pegs would not be useful for this job. Establish the words 'conductor' and 'insulator' as meaning materials that do and do not conduct electricity respectively. Encourage your class to establish fair testing conditions when trying out the materials listed on photocopiable page 22. Ask the children to work in groups and to report their results.

Answer to teacher's tea-break teaser: If the bird's going to get a nasty electric shock the electricity must flow through its body. But electricity must have somewhere else to go before it can flow – as in a circuit. So if Percy isn't touching the ground or a pylon at the same time as touching the wire then he is safe.

Page 23: The Big Switch
Learning objective
That a switch can be used to make or break a circuit. To construct circuits incorporating a range of switches.

Recap any work that your class may have done on the use of conductors in a circuit. Ask the children to imagine that every electrical appliance they knew of was turned on all day and every day. Ask them what we do in order to control these items and not waste electricity. The answer is that we stop the flow of electricity by breaking the circuit. Touch on the use of switches at this point (there is an activity on that coming up, see photocopiable page 24) but focus on breaking the circuit itself.

Use the story on photocopiable page 23 to establish the need for a switch. Put the class into goups and set each group up with a simple circuit, including a light bulb. You can cut out Kemplar's face from the photocopiable and attach it to the bulb for added impact! Set each group the task of breaking the flow of electricity without touching the battery cell. Ask the groups to record their findings and demonstrate their ideas to the class.

Page 24: Finding switches
Learning objective
That a switch can be used to make or break a circuit. To consider what sources of information, including first-hand experience, they will use to answer questions.

Start by switching your class lights on and off, asking a volunteer to do the same to a CD player. Encourage your class to identify other switches at school and at home, asking them why we need switches and how we use them. Remind your class that we never play with switches under any circumstances.

Ask the children to use photocopiable page 24 to help identify and record this information, presenting it in groups at the end of the session.

Page 25: Super switches!
Learning objective
That a switch can be used to make or break a circuit. To construct circuits incorporating a range of switches. To represent circuits by drawings and how to construct circuits on the basis of drawings and diagrams using conventional symbols.

Take a piece of wire from a circuit and ask your class what role it plays. Settle on 'it is a conductor' and explain that electricity is conducted through the wire and that this is in the form of electrons. Electrons run very quickly, as if they are in a race.

Use photocopiable page 25 to illustrate this analogy, introducing the concept of an electrical 'current' that 'flows'.

Ask the children to build a circuit from the plan on the photocopiable sheet and to test it out using class switches. Encourage them to progress to designing their own circuits and including switches they have made themselves (a small piece of cardboard with paperclips kept in place by brass paper fasteners is a good model). Ask the children to demonstrate their designs and to consider switches we use in everyday life.

PART 4:
MAKING CIRCUITS

Page 26: Series Circuits
Learning objective
To make a complete circuit using battery, wires and bulb.
That a complete circuit is needed for a device to work.
To construct circuits incorporating a battery.
Drawing circuits and constructing circuits from drawings.

Explain to your class that for electricity to flow it must have somewhere to flow to. A circuit is a wire that is arranged in a circle for the current to flow along; there may be switches and bulbs along the way.

Use your classroom lights as an example, encouraging the children to help you draw a simple circuit to illustrate how they work.

Using photocopiable page 26 to reinforce the concept of electricity flowing along a circuit, ask your class to make their own simple series circuit and record their designs.

Page 27: Parallel Circuits
Learning objective
To make a complete circuit using battery, wires and bulb.
That a complete circuit is needed for a device to work.
To explain observations in terms of knowledge about electrical circuits.
To construct circuits incorporating a battery.

Recap any work your class may have done on series circuits and introduce the idea that one switch may be used to control more than one light bulb.

Ask the children to use the circuit diagram on photocopiable page 27 to make their own circuit. You can add other features such as a switch or buzzer if appropriate. Photocopiable page 26 shows how your class could record their work – encourage the children to explain their ideas and present their findings.

Page 28: Resistance!
Learning objective
That the brightness of bulbs can be changed.
How changing the number or type of components in a series circuit can make bulbs brighter or dimmer.

Briefly recap with your class that electricity flows and that it is made up of electrons. Ask the children to work in groups, setting up a series circuit with one bulb. Call this Circuit 1. Then challenge the children to add another bulb and then another, asking them if they notice any differences regarding the brightness of the bulb.

Bulbs offer resistance to the current, so a single bulb is brighter than two and two are brighter than three. The more resistance, the dimmer the bulbs. Use photocopiable page 28 to record this experiment, and encourage the children to add more bulbs until there is too much resistance for any to light up.

Explain to your class that we use fuses to manipulate resistance for safety reasons. Split the class into groups, each with an open-backed plug. Encourage them to look closely and identify the fuse, focusing on the numbers written on the outside casing of the fuse itself.

SAFETY NOTE! Remind the children that playing with plugs at home is not to be done at any time and that the plugs you have handed out are not connected as otherwise they would be dangerous to look inside.

Page 29: Traffic lights
Learning objective
To explain observations in terms of knowledge about electrical circuits.
To construct circuits, incorporating battery or power supply and a range of switches to make electrical devices work.

Discuss with your class how electricity is used in the town or city in or near where they live. Use photocopiable page 29 as an example of people not always trusting it as they do today!

Ask the children to work in pairs to make three separate circuits to power a set of traffic lights, with one circuit for each colour light. Each pair will need three bulbs, batteries, switches, conductors and coloured cellophane.

Ask the children to present the finished product to the class and to individually write a step-by-step guide to making and operating their traffic lights.

PART 5:

MAKING ELECTRICITY

Page 30: Static electricity
Learning objective
Thinking creatively to try and explain how living and non-living things work, and to establish links between causes and effects.

Start by asking the children if anyone has ever experienced an electric shock, being careful to make it clear that giving an electric shock on purpose is not a good idea and reminding them to keep well away from sockets. If part of your classroom has carpet, it is likely to be nylon based and rubbing your shoe on it should produce an example of static electricity conducted through the body. The next thing you touch will feel a small 'crackle'.

Using photocopiable page 30 to show the lengths that people have gone to to produce static electricity, start the balloon experiment with your class, working in pairs. Use a pump to inflate the balloons, as they can be a choking hazard. It is also wise to check if any children have a latex allergy. Elicit hypotheses from the children, encouraging them to compare them to their results afterwards.

Page 31: Static movement
Learning objective
Using simple equipment appropriately.

Explain to your class that static electricity can produce movement. A machine that generates static electricity has printed the photocopiable in their hands. This attracts toner powder to form letters and shapes based on an original document. There is an excellent cartoon explanation of this on pages 44 to 46 of *Shocking Electricity*.

Use the activity on photocopiable page 31 as an experiment for your class to carry out in groups, encouraging the groups to compare hypotheses and results along the way.

Answer: b) The atoms of the clingfilm are short of electrons. This means they are positively charged and give out positive forces. Remember how two negative forces push each other away? Well, two positive forces also push against each other and that's why the two pieces of clingfilm move apart. The comb rips electrons off your hairs and the force from these electrons (negative charge) pulls in the positively charged atoms in the clingfilm.

Page 32: Making lightning
Learning objective
Thinking about what might happen.

Start by asking your class about their experiences with lightning: what they know about it, how it can look and sound etc. Use examples from books and TV. (Allan Ahlberg's *It was a dark and stormy night*, Scooby Doo may help.)

Explain that lightning can be dangerous but when precautions are taken we are much safer. Lightning conductors are used on large buildings, most shoe soles contain rubber and the chances of being hit are very slight.

Darken your room as much as possible and use the activity on photocopiable page 32 to make the link between a build-up of static electricity in clouds and the formation of lightning. Encourage the children to add to the fact file with an account of their experiment and to draw what they did in strip cartoon form for a class display.

Answer: c) These are electrons stripped off the wool jumping from the balloon to the radio aerial. The sparks are basically tiny flashes of lightning.

Page 33: Hearing lightning
Learning objective
Asking questions that can be answered scientifically.

If you have a tape of lightning sound effects, use this as a starting point. Encourage the children to describe in writing how it sounds to them.

Use photocopiable page 33 to make and hear lightning, and contrast this to the sound on the tape, writing these descriptions too. Use 'The shocking details' information from photocopiable page 32 as a starting point to encourage the children to research this topic in groups.

Answer b) You are listening to electrons jumping from the balloon to the radio aerial. If you switch on the radio during a thunderstorm tuned as before you will hear the same noise but this time it will be made by lightning. Mind you, you won't have been the first to investigate this sizzling force of nature.

PART 6:
THE AWFULLY BIG QUIZ

Page 34: The Awfully Big Quiz
Assessment:
Constructing circuits incorporating a range of switches.
How changing the number or type of components in a series can make bulbs brighter or dimmer.
Representing circuits on the basis of drawings and diagrams using conventional symbols.

Ask your class to tell you what they have learned about electricity, encouraging them to point out the 'best bits' of the topic. Explain that you are going to have a quiz show in class, and divide the children into teams. Encourage the children to think of an electrically themed name for their team, using photocopiable page 34 to focus them on the first team task: making their buzzer for the quiz itself. Mark the teams on how well they work as a unit as well as their final design.

Page 35: The Awfully Big Quiz 2
Assessment:
Constructing circuits incorporating a range of switches.
How changing the number or type of components in a series can make bulbs brighter or dimmer.
Representing circuits on the basis of drawings and diagrams using conventional symbols.

Use the questions on photocopiable page 35 as the basis for the first round of your quiz, encouraging the children to research questions from their own science work in class to add more questions for opposing teams.

Page 36: The Awfully Big Quiz 3
Assessment:
Constructing circuits incorporating a range of switches.
How changing the number or type of components in a series can make bulbs brighter or dimmer.
Representing circuits on the basis of drawings and diagrams using conventional symbols.

WHAT'S THIS?

Use photocopiable page 36 as the basis for a multiple choice round of your quiz. Encourage the children to research their own examples and to make up bogus, yet plausible, answers to try and outwit their opponents.

Answers: 1 a) Even if the radio doesn't work off mains electricity it will be powered by electricity from a battery. When you are chatting to a friend on the phone the receiver turns your voice into electrical signals that travel down the wire to your friend's phone where they are turned into sounds again. Toilets aren't powered by electricity but you may be interested to know that in 1966 inventor Thomas J Bayard devised an electrically powered wobbling toilet seat. The idea was that pummelling the bum prevents constipation. Sadly, people poo-poohed the idea and the seat went off the market.

2 b) This is handy because the person is usually thrown a safe distance from the object that is giving them the shock. Another effect of a violent shock to the muscles is to make you poo and pee... resulting in shockingly smelly underwear.

3 b) Electricity can pass through water – which is why it is extremely silly to put any electrical machine near water or to touch power sockets or switches with wet fingers.

Page 37: The Awfully Big Quiz 4
Assessment:
Constructing circuits incorporating a range of switches.
How changing the number or type of components in a series can make bulbs brighter or dimmer.
Representing circuits on the basis of drawings and diagrams using conventional symbols.

Explain who Benjamin Franklin was and talk about his importance in the story of electricity.

Encourage the children to add more visually based questions to challenge the opposing teams. You could award points for well-researched questions as well as correct answers.

Answers: 1 b) RIVER. The Thames was full of dead cats, rats and floating poo so swimming in it was almost like committing sewer-cide. Franklin also enjoyed writing letters in the bath (no doubt when he was trying to get clean afterwards).
2 c) ROCKING CHAIR. Does that make Franklin a rolling rock star?
3 e) FARTS. The competition might have been nothing to sniff at but it failed to find a winner.
4 a) TORNADO. He hit the tornado with his whip but

it just whipped around as usual.
5 d) TURKEY. Do you reckon the USA would be such a powerful nation today if its national symbol was a turkey?

PART 7:
LITERACY & MATHS

Pages 38 & 39: Wordsearch Clues and Wordsearch Puzzle
Learning objective
Contextual understanding.

Encourage your class to make their own wordsearches on squared paper after trying out this example. The children's self-written puzzles could be used in a class book for visitors to try for themselves.

Page 40, 41 & 42: Straight from the Heart 1, 2 & 3
Learning objective
Create, adapt and sustain different roles in drama.
Discuss and evaluate their own and others' writing.

Explain to the children that this is a true story. After reading it aloud as a class, focus on the plot and ask the children what happens next in order to establish a clear sequence of events. Divide these events up into scenes using the prompts on photocopiable page 42.

Recap any playwriting or performing that your class may have done and link these to soap operas on TV. Although the characters and events on screen seem real, the actors are really remembering dialogue from their scripts and doing what the director decided is best. Use the start of the script on photocopiable page 42 to show how these scripts are set out, asking for volunteers to read out the different parts.

Ask your class to look again at the story *Straight from the Heart*. Encourage them to list the plot points for each scene. Once this is established the children can then start on their scripts.

Invite the children to share their work in small groups or pairs and to review and improve what they have written. Encourage small performances and rehearsals in class. If possible, you could extend this activity to a class assembly or performance, using scenes written by different children in a final photocopied class script.

Page 43: Straight from the Heart 4
Assessment:
Simple circuits.
Everyday effects of light, vibration and sound.

Use photocopiable page 43 to encourage your class to design a poster for their dramatic adaptation of *Straight from the Heart*. Discuss the features of a good poster and show any examples you may have. Decide as a class what information needs to be included and any ideas that the children may have. Remind them that the poster must be eye-catching to interest any potential audience members.

Split your class into teams, one to design a circuit to illuminate the poster, another to build it and another to test it out. Each team should report back to the class appropriately.

Once completed, invite your headteacher to switch on the poster in an opening ceremony, asking each team to give their final reports detailing how it was produced.

Page 44: Lightning Addition Quiz
Learning objective
Choose, use and combine any of the four number operations to solve word problems.

Use photocopiable page 44 as a quick quiz for individuals or pairs using pencil and paper methods from their numeracy work in class. Encourage the children to add their own questions for others to try out and to come up with their own wacky inventions for a class display.

Answers: 1) 100.
2) 7. US Park ranger Roy Sullivan was struck *seven* times between 1942 and 1977.
3) 5.5. Lightning can be 30,000 degrees Celsius - the surface of the sun is a cooler 5,530 degrees Celsius.
4) 17. In 1995 children and parents at a football match were struck by lightning in Kent, England. They all survived, but some had nasty burns.

PART 8:
RESEARCH

Page 45: Shocking Fact File
Learning objective

Consider what sources of information to answer questions.

Split your class into research teams, using photocopiable page 45 as a starting point. Encourage the children to use the Internet and school library, as well as continuing their research at home. Award research prizes for particularly interesting or entertaining examples and praise well-ordered research and notes. Add each team's efforts together into a class book of Shocking Facts.

Page 46: Shocking Inventions 1
Learning objective
That science is about thinking creatively.

Start by telling the children that many people work on new inventions all the time and that some are more successful than others. Use photocopiable page 46 as a starting point for the children to consider the pros and cons of this particular real invention. Encourage the children to use their electrical knowledge to write a lightning safety guide that may prove more useful than the invention itself!

Page 47: Shocking Inventions 2
Learning objective
That science is about thinking creatively.

Recap the many different electrical items that we use every day and the different jobs that they do. Use photocopiable page 47 to focus the children on what makes an invention successful. Talk about the practicality of an electric-powered car and its pros and cons. Ask the children to think of other electrical items and extend this to possible uses of electricity in the future as the world's natural resources begin to run out. Encourage the children to write a catalogue of the future, giving information about electrical items in the late 21st century.

Page 48: My cover
Learning objective
To design and make images.

Use photocopiable page 48 as a template for your class to design their own individual science book covers for this topic.

NAME _____   DATE _____

Our electricity

- Imagine that your classroom has no electricity – just like Horra Island!

- Make a list of everything you can see that uses electricity.

My list

- Put a circle around anything in your list that you can power in a different way or replace with something that does not need electricity.

- What do you use electricity for at home? Draw a picture of a room at home and label it to show as many different electrical items as you can.

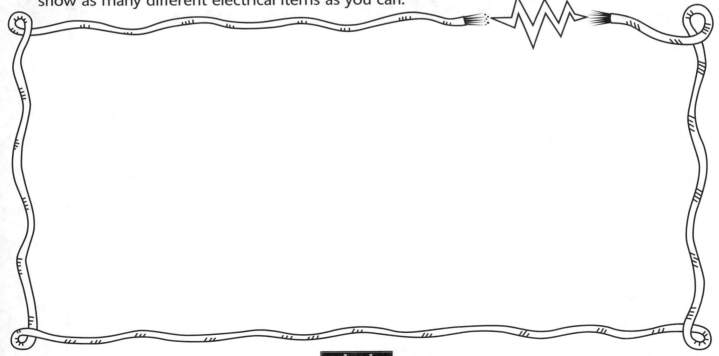

NAME _____ DATE _____

Mains and Battery

- Horra Towers needs electricity!

- Some of the electrical items in the children's letter use mains and some use batteries.

- Can you help the children on Horra Towers electrify the island?

- Use the data file below to list the items they mention in the letter. Then decide whether these items use batteries or mains electricity.

- Can you think of any other electrical items that the children might use?

- Add them to the data file and decide whether they use batteries or mains electricity.

Horra Towers
Horra

Dear Coastal Rescue
Please rescue us from Horrible Horra!
This island has NO ELECTRICITY and no electric heaters. It's FREEZING cold and we're taking turns to warm ourselves on the island's cat. All we've got to eat is cat food because our food supplies have been lost. And it's not even hot cat food because there aren't any electric cookers.

Our only light is a smelly candle – 'cos light bulbs need electricity. And it's MEGA-BORING here 'cos there's no TV, no videos, no computer games, and no CD player 'cos, yeah, you guessed it – they all need ELECTRICITY. And our teacher, Mr Sparks, is making us do extra homework. He says as a reward we can listen to him playing his squeaky old mouth organ.
Laugh, we nearly cried.
You've got to come before we all die!!!! Pleeeeeeease!
Lots of love,
Class 5e

PS The cat would like some fresh fish for supper.

Item	Battery	Mains

Power cut!

- Take a look at this cartoon.

- Imagine that the electricity still wasn't working the next day!

- Make some notes about how this would affect your life.

NAME —————————————— DATE ——————————

THE POWER CUT CONTINUES

● Continue the cartoon to tell how the power cut would affect your life
if it carried on the next morning!

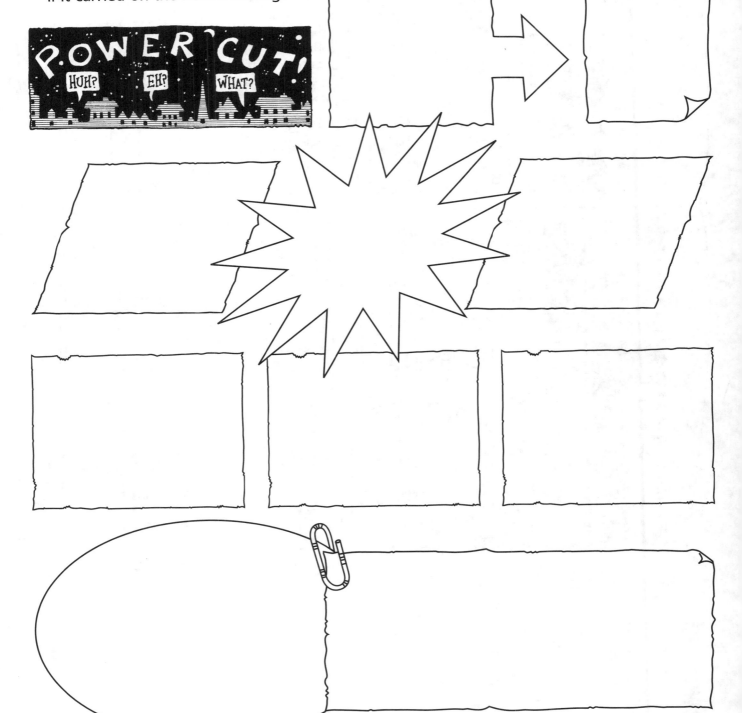

NAME _____   DATE _____

Safety First!

● Humans have discovered that electricity has many uses – but sometimes we get it wrong!

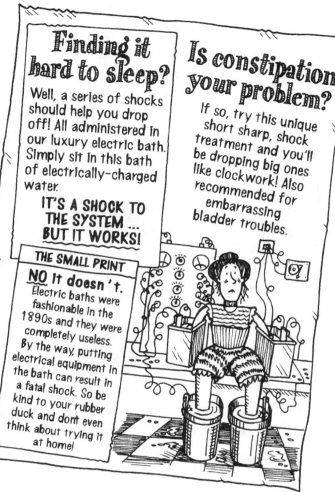

● Not only are these inventions useless but they could also be dangerous!

● Use this plan to design a safety poster telling people how to use electricity responsibly.

My poster

Slogan: _____

Text:

_____
_____
_____
_____

Drawing:

● Now you can draw your poster – remember to make it eye-catching!

Spot the electric motor!

Spot the electric motor

Which of these household objects contains an electric motor? (No, you're NOT allowed to take them apart to find out.) Here's a clue instead – if it's got moving parts it's got an electric motor.

- Electric motors are used to power many different items that do different jobs.

- Electricity can be used to heat, cool, make a sound and light.

- Use the data file below to classify the items on this sheet according to what their main job is, then add some more of your own. One has been done for you.

Appliance	Mains or battery?	Location	Heat	Cool	Light	Sound
Kettle	Mains	Kitchen	yes	no	no	no

NAME _____ DATE _____

How to make an electric motor!

● Follow these instructions to make an electric motor.

What you need is:
A compass OR a needle
A magnet
A 25 cm (10 inch) length of thread
some blutak
sticky tape
A 1.5 volt battery (HP11)
A piece of kitchen foil 28 cm x 6 cm (11 x 2.3 inches)
A grown-up to help. (Yes, they have their uses.)

What you do is:
1 If you don't have a compass, stroke the needle with the magnet 30 times. This turns the needle into a magnet too.
2 Secure the needle to the end of the thread with a small blob of blutak in the middle so it hangs sideways in the air.

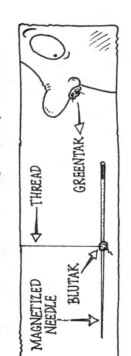

MAGNETIZED NEEDLE
BLUTAK
THREAD
GREENTAK

3 Stick the other end of the thread to a table top with more blutak.

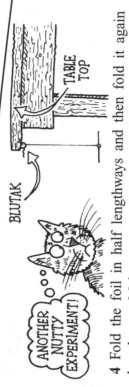

ANOTHER NUTTY EXPERIMENT!

BLUTAK
TABLE TOP

4 Fold the foil in half lengthways and then fold it again lengthways. Make sure you don't tear the foil.
5 Use sticky tape to stick one end of the foil to the positive end of the battery and the other end to the negative end. This makes a circuit for an electric current to run through.
6 Now, EITHER ... hold the battery horizontal and pass the foil loop from side to side over the face of the compass. OR put the foil loop round the needle and move the foil up and down without touching it.

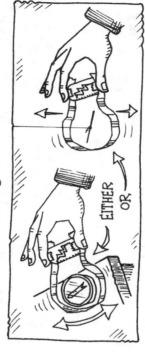

EITHER OR,

What do you notice?
a) The needle starts to glow with a strange blue light.
b) The needle twists round.
c) The needle jumps up and down.

● Record your observations here.

_____
_____

How to make an electro-magnet!

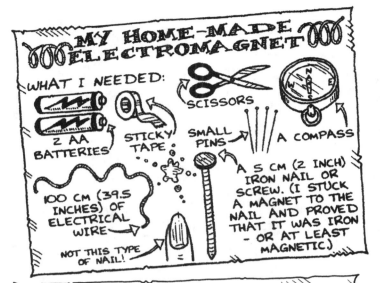

MY HOME-MADE ELECTROMAGNET

WHAT I NEEDED:

2 AA BATTERIES

STICKY TAPE

SCISSORS

SMALL PINS

A COMPASS

100 CM (39.5 INCHES) OF ELECTRICAL WIRE

NOT THIS TYPE OF NAIL!

A 5 CM (2 INCH) IRON NAIL OR SCREW. (I STUCK A MAGNET TO THE NAIL AND PROVED THAT IT WAS IRON – OR AT LEAST MAGNETIC.)

- Electric motors use electro-magnets.

- Magnetism and electricity are similar as they both move atoms.

- Follow these instructions to make an electro-magnet of your own!

WHAT I DID:
1 I sticky-taped the ends of the two batteries together like so.

POSITIVE END

STICK THE POSITIVE TO THE NEGATIVE END

NEGATIVE END

2 Then I stripped the plastic wrapping from the ends of the wire – 0.5 cm (0.15 inches) at each end – to expose the copper wire inside.

⚠ **HORRIBLE DANGER WARNING!**

This requires a sharp knife. You *must* recruit an adult for wire-stripping or you might feel a bit cut up.

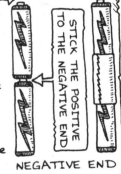

3 I used more sticky tape to stick one end of the wire to the positive end of the battery.

4 Next I wrapped the wire tightly around the entire length of the nail – and wrapped it around the length of the nail twice more.

5 Now I stuck the other end of the wire to the negative end of the battery.

6 I moved the wire backwards and forwards next to the compass.

RESULT:
Aha – just as I expected! The compass needle started to swing backwards and forwards without me touching it.

SWING!

REMARKS:

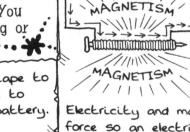

MAGNETISM

MAGNETISM

By joining the wire to each end of the batteries I made what we scientists call a circuit for the electricity to run along. (It flows from the negative to the positive end.)

Electricity and magnetism are the same force so an electric current also gives off magnetism. The tightly wrapped wire meant that the magnetic force was strong enough to affect the magnetic needle in the compass. The magnetic force was also strong enough to pick up the small pins. How satisfying!

- Well done!

- Now draw a diagram of your electro-magnet in action!

NAME _____ DATE _____

THE BATTERY

The battery cell was invented in 1865 by French inventor Georges Leclanche. It uses a mixture of chemicals to make chemical reactions that result in electrons flowing from the zinc inner container to the carbon rod.

● Take a look inside your torch.

● Draw what you find. Include wires, springs, bulb, and of course – the battery cells!

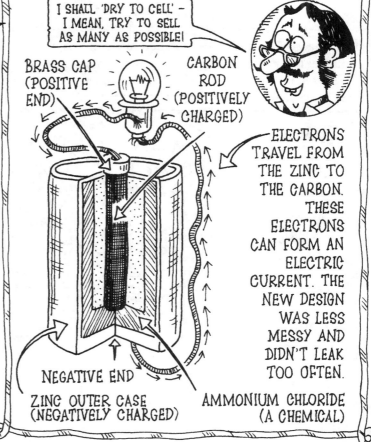

I SHALL 'DRY TO CELL' – I MEAN, TRY TO SELL AS MANY AS POSSIBLE!

BRASS CAP (POSITIVE END)

CARBON ROD (POSITIVELY CHARGED)

ELECTRONS TRAVEL FROM THE ZINC TO THE CARBON. THESE ELECTRONS CAN FORM AN ELECTRIC CURRENT. THE NEW DESIGN WAS LESS MESSY AND DIDN'T LEAK TOO OFTEN.

NEGATIVE END

ZINC OUTER CASE (NEGATIVELY CHARGED)

AMMONIUM CHLORIDE (A CHEMICAL)

Inside my torch:

Could you be a scientist?
Which way round do the batteries go in a torch? Yeah, OK you can try it out – or you could think about which way the electrons move. Is it…

a) Positively charged end to positively charged end? ☐

b) Negatively charged end to negatively charged end? ☐

c) Negatively charged end to positively charged end? ☐

● Clue: the symbols that you use in maths will help you!

NAME _____   DATE _____

The Light Bulb

A modern light bulb uses a coiled tungsten metal filament surrounded by argon – a harmless gas found in the air.

FILAMENT

THE ARGON GAS ACTS AS A BUILT IN FIRE EXTINGUISHER

ARGH! I'M 'ARGON'ER

FLUTTER

● Take a closer look at your light bulb.

● Now draw a diagram of it in the box below.

My bulb:

● See if you can add these labels to the right parts of your bulb:

Filament

Glass

Base

Metal screw thread

Bet you never knew!
Light bulbs save lives. Nowadays many lighthouses use electric light from powerful bulbs to warn ships away from rocks. Each lighthouse has its own pattern of flashes so that sailors can work out where they are in the dark.

ER, HANG ON ...TWO LONG FLASHES, ONE SHORT ONE...

CRUNCH!

● What other ways can light bulbs help us if we are in danger? Make a list with a friend.

NAME _____   DATE _____

THE DAILY SUN
31 December 1879

Blaze of glory!

Heroic inventor Thomas A. Edison glowed with pride as he showed off his new invention.

Thousands of people watched him light up the whole town with 3,000 newly-invented light bulbs. Each one looks like a globe of sunshine. No more will people huddle in the dark afraid of the shadows. Thomas Edison is a shining example to the whole nation! It's such a pity that 14 of the new light bulbs have already been pinched.

YOU, TOO, CAN OWN YOUR VERY OWN LIGHT BULB. $50 each

THE SMALL PRINT
You need to have electricity connected to your home to make them work. Mr Edison hopes to have an electricity supply system up and running in the next year or two.

NOTICE TO THE PUBLIC
To light one of Mr Edison's lights you simply have to flick a switch. Do not try to light them with matches.

Selling Light Bulbs!

- Before electric light bulbs, most people used candles and paraffin lamps.

- Some wealthy people used gas light but it was smoky and smelly – and sometimes it exploded!

- Look at the advert below the story in *The Daily Sun*. It is trying to persuade people to buy electric light bulbs.

- See if you can design a pamphlet to tell everyone about this fantastic new invention. Use the plan below to help you.

My plan:

Cover (including slogan):

Introduction:

How light bulbs work:

Where you can use them:

Why they are safer than gas or paraffin:

Safety points about light bulbs:

Diagrams and other pictures I want to include:

NAME _____   DATE _____

Conductors

● Set up your circuit to include a bulb, battery cell and wires with clips at either end. Make sure that the bulb is lit.

● Now unclip one of the wires.

● What do you notice happens to the light bulb? Draw a diagram showing how your circuit looked before you unclipped the wire and another showing what it looks like now.

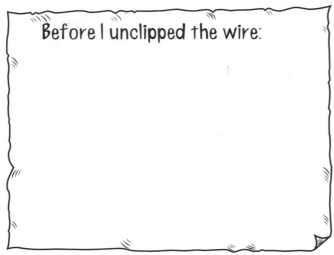

Before I unclipped the wire:

Now:

● Use the items in the list below to complete your circuit. Some will conduct electricity, others will not. They are called insulators. How will you tell if the object is a conductor or an insulator?

Object	My prediction	Result	Conductor or insulator?
Wooden ruler			
Lead pencil			
Felt-tipped pen			
Metal spoon			
Paper strip			

Teacher's tea-break teaser
You will need a bird. No, not a real pigeon like Percy – a toy bird will do. All you do is tap gently on the staffroom door. When it opens, smile innocently and ask:

HOW COME BIRDS CAN PERCH ON A HIGH-VOLTAGE WIRE AND NOT GET ELECTROCUTED?

NAME _____   DATE _____

THE BIG SWITCH

- George Westinghouse was a rich businessman who spent a lot of money developing electricity in the USA.

- Unfortunately he was not always happy with the way electricity was used...

- Can you help George Westinghouse stop the execution?

- Set up a circuit so that the bulb is lit.

- Can you work out a way to turn the bulb off?

- Record your findings below.

NEW YORK NEWS
December 1888

A SHOCKING WAY TO GO!

New York State is to execute murderers. This follows embarrassing incidents when hangings have gone wrong and people have had their heads pulled off by the rope. The first victim to be

electrocuted is to be William Kemplar a fruitseller convicted of killing his girlfriend. Thomas Edison says the execution will prove the danger of alternating current.

W. Kemplar (before the murder)

⚡ STOP PRESS! ⚡

Westinghouse is shocked that his alternating current is going to be used to kill someone. Kemplar says he's shocked too and he's appealing, claiming the execution is too cruel. We expect him to be even more shocked if the execution goes ahead.

W. Kemplar (Yesterday)

Things we used (equipment):

What we did (method):

What we thought might happen (hypothesis):

What happened (results):

What I learned (conclusion):

NAME _____ DATE _____

Finding switches

● We use switches to control the flow of electricity.

● Look at the electrical items in your classroom.

● Can you find the switches?

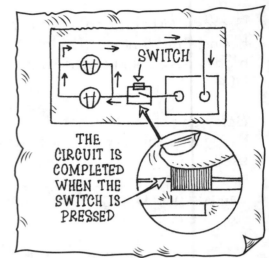

THE CIRCUIT IS COMPLETED WHEN THE SWITCH IS PRESSED

● Draw the items in the box below and label the switches. Explain what the item does when it is switched on and how you know that it is switched off.

● Sometimes younger children play around with electrical switches. Write a sentence that will explain to them why we *never* play with electricity.

_____

NAME _____ DATE _____

Super switches!

- Electricity is made up of electrons that flow along the wires of a circuit.

- These electrons are really being put to the test!

Are you ready to make the switch? The electrons sure better be! In this exercise they'll have to get past the dreaded electrical switch. The switch is a springy piece of metal. When the switch is on the springy piece of metal is held down so the electrons crawling through the wire can crawl through it too. But they'd better be quick because when the switch is off the metal springs up and breaks the circuit. Leaving the electrons stranded!

THE CIRCUIT IS COMPLETED WHEN THE SWITCH IS PRESSED

- Can you build the circuit in the diagram? Put it to the test to see if it works.

- Now design and build your own switch circuit and see if the electrons can manage it! Use the planning box for your design.

My plan:

NAME _____ DATE _____

Series Circuits

● Imagine the electrons are at a training camp...

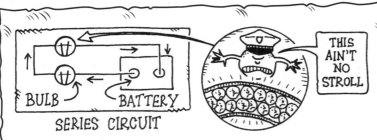

The electricity in your sockets is actually made up of moving electrons and the power of electricity comes from the force the electrons give out.

The first race is the series circuit – it's a nice, gentle warm up. The electrons must run from their battery hut round the wire, through the bulbs, and back to their battery.

● See if you can make the series circuit to test the electrons' fitness.

● Use these headings to plan and carry out your experiment.

What we used (equipment):

What we did (method):

What I thought might happen (hypothesis):

What happened (results):

What I learned (conclusion):

● Now see if you can design your own series circuit using three bulbs to test the electrons even more! Draw your design in the planning box below.

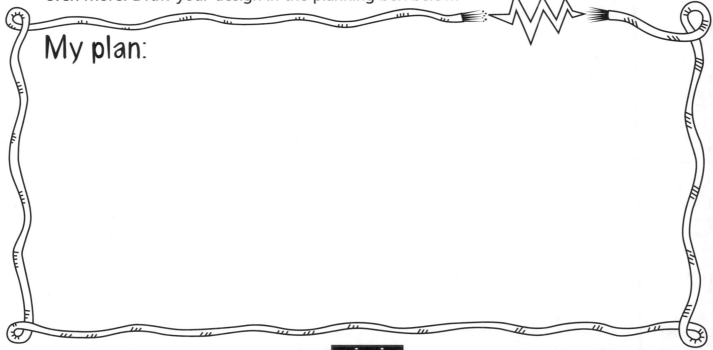

My plan:

NAME _____   DATE _____

Parallel Circuits

- Electrons move around a circuit to power a light bulb. Imagine it as a race at a training camp.

- A.Tomm is in charge of training! He makes some races harder than others. The series circuit was easy compared to this...

A.Tomm has rearranged the wires so that there are two separate wires for each bulb. This means half the electrons can go one way and half can go the other.

- This is called a parallel circuit. Can you make it?

- Now see if you can design your own parallel circuit to test the electrons even more! Draw your design in the planning box below.

My plan:

NAME _____   DATE _____

Resistance!

● Electrons flow through wires to power the bulb. Imagine they are in a race...

FUSE

GO FOR THE BURN, KIDS!

The electrons have to crawl through a narrow piece of wire. The resistance they get as they crawl through makes a lot of heat.

If too many crawl in together the wire may melt so it's a real dangerous work-out.

Circuit 1	Circuit 2	Circuit 3

● Draw Circuits 1, 2 and 3 in these boxes, adding a light bulb each time.

● What do you notice happens to the bulbs?

● Why do you think this is?

● Draw the inside of a plug. Where is the fuse?

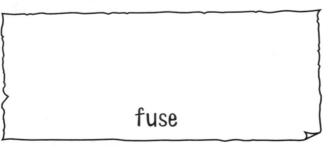

fuse

plug

● Now draw the fuse close up.

● How much current will it allow through? How can you tell?

NAME _____   DATE _____

TRAFFIC LIGHTS

● Nowadays all towns and cities in the UK use electricity. Electricity even helps us get around them safely!

● Do you know how traffic lights work? Set up three separate series circuits with switches to operate the red, amber and green lights separately.

● Remember, the sequence should go: red, red/amber, green, amber.

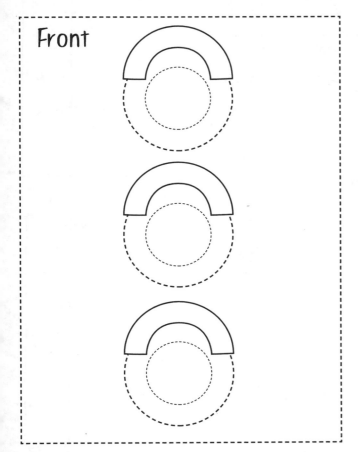

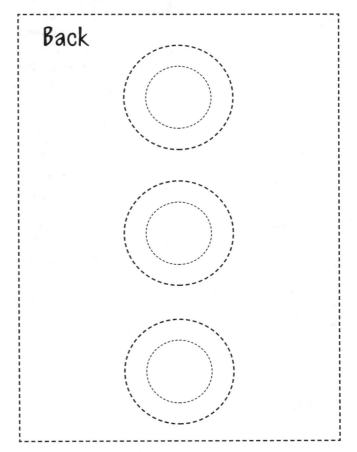

● Cut out the inner circles for the bulbs to shine through and add coloured cellophane or tissue paper.

● Now see if you can operate the traffic lights!

NAME _____   DATE _____

Static electricity

Bet you never knew!
Ancient Greek boffin Thales of Miletus (624-545 BC) made static electricity by rubbing amber (a kind of fossil tree gum) with an old bit of fur (I hate to think what happened to his pet cat). The amber could then pick up feathers.

HE USED TO PURR WHEN I DID THIS

BIT OF OLD FUR

STROKE!

AMBER

Well, if that's sparked your interest maybe you'd like to try that experiment too (hopefully your cat will manage to keep her fur on).

● There is an easier way!

● Equipment: You will need

A balloon   Piece of paper   Woollen jumper

● Method: This is what you do

Rub the inflated balloon quickly over the jumper ten times.

Place it on top of the paper.

● Hypothesis: What do you think will happen?

● Results: What actually happened?

● Draw a strip cartoon for Thales of Miletus explaining how you carried out your experiment. Use captions to include the information he will find useful.

NAME _____   DATE _____

Static movement

What you need is:
Two pieces of new clingfilm 10 cm x 2 cm (4 inches x 0.75 inches).
A clean dry comb
blutak
Some clean hair – you might possibly find some on your head. (If not maybe you could ask the cat nicely.)

What you do is:
1 Hold a piece of clingfilm in each hand. Try to bring the two pieces of clingfilm together. Notice what happens.
2 Stick a piece of clingfilm to the end of a table so the clingfilm hangs downwards. Now comb your hair quickly and strongly four times. Quickly point the teeth of the comb towards the strip of clingfilm and hold it close but not touching.

● What do you think will happen? Write your hypothesis here.

Well, what does happen?
a) The two pieces of clingfilm are drawn together. But the clingfilm doesn't want to touch the comb.
b) The pieces of clingfilm don't want to touch but the clingfilm does want to touch the comb.
c) A spark flies between the clingfilm and comb but nothing happens between the two bits of clingfilm.

● Write a sentence describing what happened in more detail.

NAME _____ DATE _____

Making lightning

- Describe what you see in your own words.

_____
_____
_____
_____
_____

THE SHOCKING DETAILS:
1 A bolt of lightning strikes at 1,600 km (1,000 miles) a second. BOOM!

2 Lightning can flash inside the cloud from bottom to top. This is called sheet lightning. CRACK!

3 Lightning can strike the ground or even leap upwards from positively charged atoms on the ground. This lightning has more energy and moves at 140,000 km (87,000 miles) a second!

What you need is:
A radio with the aerial extended
A balloon
A thick jumper (no not a stupid kangaroo – I mean a *woollen jumper*). A woollen rug or scarf will also do.

What you do is:
1 Wait until it gets dark or sit in the coal cellar with the lights out. This experiment works best in complete darkness.

ERK! OOF! EEK! OUCH! MIND THE STEP!

2 Rub the balloon on the wool about ten times. Put it near or touching the aerial.

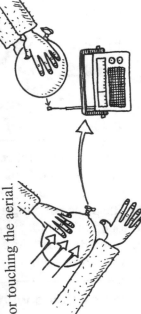

What do you see?
a) The radio comes on without you touching it – it's spooky.
b) An eerie glowing ball of light appears and floats round the room – scaring the life out of your pet budgie.
c) Tiny sparks.

NAME _____   DATE _____

Hearing lightning

What you need is:
The same equipment from 'Making lightning'.

What you do is:
1 Switch the radio to AM and make sure it's not tuned to any station.
2 Turn the volume down very low.
3 Repeat the 'Making lightning' experiment and listen.

HMMM!

What do you hear?
a) Pop music even though the radio isn't tuned.
b) A quiet pop (but it isn't music).
c) You hear the sound in **b)** but it's REALLY LOUD.

● Describe what you hear in your own words.

_____
_____
_____
_____
_____
_____
_____
_____
_____
_____
_____
_____
_____
_____
_____

Teacher's tea-break teaser
Try this shockingly tricky question on your teacher...

HOW CAN WATER START A FIRE?

STAFF

Clue: it's to do with static electricity.

Answer:
Sometimes when oil tanker holds are being cleaned with high pressure hoses, atoms in the water rub together really fast. This makes static electricity that results in lightning sparks. The sparks can set fire to petrol fumes in the hold and blow up the tanker!

● What other facts can you discover about lightning? Use books and the Internet to help you start a lightning data file!

NAME _____   DATE _____

The Awfully Big Quiz

WHAT'S THIS?

ST

- To play the Awfully Big Quiz each team will need a buzzer.

- You will need:

    Battery cell   Buzzer

    Switch   Wires

- Design and label a series circuit to include a buzzer in the box.

- Now build your circuit and test it out.

Our design

What do you think will happen when the switch is on?

_____

_____

Why does this happen?

_____

_____

- Write or draw other uses for buzzers here.

NAME _____   DATE _____

The Awfully Big Quiz 2

● Cut out these quiz cards.

● Some have been left blank for you to research questions for your opponents to answer.

HORRIBLE SCIENCE	HORRIBLE SCIENCE	HORRIBLE SCIENCE	HORRIBLE SCIENCE
Q: Name three things in this room that use electricity.	Q: Are TV sets usually mains or battery powered? **Bonus**: Two points if you can explain why!	Q: Name three electrical items that we use to make sound.	Q
HORRIBLE SCIENCE	**HORRIBLE SCIENCE**	**HORRIBLE SCIENCE**	**HORRIBLE SCIENCE**
Q: Two points for each safety rule about electricity that you can remember.	True or False?:	Q: Name three items that are usually battery operated.	Q: A switch can make a battery more powerful. True or False?
HORRIBLE SCIENCE	**HORRIBLE SCIENCE**	**HORRIBLE SCIENCE**	**HORRIBLE SCIENCE**
Q:	Q: Dismantle a torch. Two points for each part you can name. **Bonus**: 4 points if you can put it together again and make it light!	Q:	Q:
HORRIBLE SCIENCE	**HORRIBLE SCIENCE**	**HORRIBLE SCIENCE**	**HORRIBLE SCIENCE**
Q Name five items *not* in this room that use electricity. **Bonus**: Sort them into mains and battery powered for two extra points.	Q:	Q: The proper word for a single battery is 'cell'. True or False?	Q:

NAME _____   DATE _____

The Awfully Big Quiz 3

- Use these questions for your quiz.

- Add some more questions here. Give three possible answers for your opponents to choose from.

1 Which of these machines doesn't need electricity to work?
a) The toilet
b) The telephone
c) The radio

2 Why is it that the victim of a huge electric shock gets thrown through the air? (No need to test this on family pets or frail elderly teachers.)
a) The force of the electricity lifts them off the ground.
b) The electric current runs through the nerves and makes the muscles jerk violently so the victim leaps backwards.
c) Electricity reverses the force of gravity and makes the body weightless for a second.

3 Your teacher gets struck by lightning in the playground during a storm. Why is it dangerous to be in the playground at the same time?
a) You might have to give your teacher the kiss of life.

NOT ME!   NO WAY!   FORGET IT!

b) The playground will be wet from rain. The electric current from the lightning can spread through the wet surface and give you a nasty shock.
c) The hot lightning turns playground puddles into dangerous super-heated steam.

The Awfully Big Quiz 4

Benjamin Franklin's puzzling pictures quiz
In this quiz you're shown pictures of the answers but they're muddled up and you'll have to match the answers to the questions correctly. Here's an example question to show you how it's done.

QUESTION
Franklin thought it was healthy to sit by an open window in winter without any of his...?

DESCRIPTION

Answer: Clothing.
(Don't try this at home. It's not very comfortable and it's probably against the law!)

JUST TO MAKE THE QUIZ A LITTLE BIT HARDER WE'VE MUDDLED UP THE ORDER OF THE DESCRIPTIONS!

QUESTIONS
1 When Franklin lived in London he enjoyed a dip in the local...?
2 One of Franklin's most popular inventions was a...?
3 Franklin organised a competition to develop a food that would result in a sweet smelling...?
4 In 1755 Franklin got on his horse and chased a...?
5 Franklin was a key figure in the struggle to make the USA independent of Britain. He suggested the American symbol should be a...?

PICTURES

a) TORNADO
b) RIVER
c) ROCKING CHAIR
d) TURKEY
e) FART

DON'T FORGET TO WORK OUT THE ANSWERS FROM THE PICTURES!

● Now make up some picture questions for the opposing teams!

WORDSEARCH CLUES

● These clues tell you fascinating facts about electricity.

● Look out for the words written in CAPITAL LETTERS. These are the words you will be looking for on the wordsearch itself.

1 In 1999 2,000 British rail travellers and 14 trains were held up by a metal foil yoghurt POT LID. The lid had got stuck in a crack on the electrical rail and diverted the electric current into the ground. Of course, without power the trains were stuck.

2 The PAUL TRAP is an arrangement of electrical and magnetic forces that traps atoms between them so that they can't get lost and can be studied at leisure. It was invented by German scientist Wolfgang Paul in 1989.

3 In 1750 physicist William Watson performed an experiment. He made a line of people hold hands linking an electricity-making machine with a CANNON. An electric current passed along the line and everyone got an electric shock.

4 During a sandstorm electric force is made by bits of SAND rubbing together. One German explorer caught in a sandstorm wore a CAR JACK (a tool for lifting up cars) on his head with a lead to the ground. The jack attracted the electric force and the lead diverted it to the ground – just like a lightning conductor.

5 One of the earliest electric inventions was meant to catch the Loch Ness MONSTER. It was designed to electrify the water and kill the poor beastie, but luckily this cruel invention never got off the drawing board. Nessie lives!

6 Lightning is a giant electric spark made by thunderstorms. A bolt of lightning can actually make fertilizer – the HEAT of the lightning causes a chemical reaction in the soil that makes NITRIC ACID. Further reactions turn the acid into chemicals called nitrates that plants need so they can grow.

NAME _____   DATE _____

Wordsearch Puzzle

● Use your clue sheet to help you find the words in this wordsearch.

● You get one point for each word, there are 12 altogether.

● The words can appear forwards, backwards, top to bottom or bottom to top.

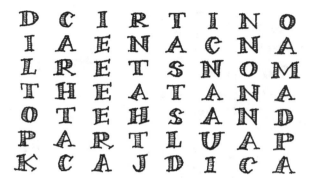

```
D  C  I  R  T  I  N  O
I  A  E  N  A  C  N  A
L  R  E  T  S  N  O  M
T  H  E  A  T  A  N  A
O  T  E  H  S  A  N  D
P  A  R  T  L  U  A  P
K  C  A  J  D  I  C  A
```

● How many did you get?

_____
   12

Bonus!

● Make your own electricity wordsearch.

● You will need a 10 x 10 square grid.

● Think of interesting clues and test your friends.

Straight from the Heart 1

Straight from the heart

Arizona, USA, 1947

"Here we have an interesting case. A 14-year-old male with a chest that hasn't grown properly for several years – making him unable to breathe normally. Am I going too fast?"

Top surgeon Claude Beck glanced at the medical students who were taking notes and, as usual, following his morning ward round like a flock of white-coated gulls after a fishing boat.

Beck had short greying hair, a square face, a square jaw and looked you squarely in the eye even when he had bad news to announce. And right now he was gazing straight into the eyes of his young patient.

"I wish I could tell you the op will be a cinch son, but it's an involved procedure. We've got to separate your ribs from your breastbone so you can breathe normally. Still I reckon we'll pull it off." Mickey's eyes were huge and dark and the rest of him looked thin and pale under his crew cut.

"And then?" he whispered anxiously.

"You'll be right as rain."

Mickey struggled to ask another question but he was short of breath and the surgeon and his students had already moved on. So later he asked a nurse about Dr Beck.

"Oh yes, Mickey," she smiled. "He's a real expert. Why, he's so clever he's even gone and developed a machine to re-start hearts using electric shocks. It's called a defibrillator. He's been testing it out on dogs. So don't you worry – you're in good hands."

Beck did indeed pull it off. The operation went just fine and after two hours the ribs were separated. The tricky part was over and the surgeon sighed with relief as he carefully sewed up the wound. Then, without warning Mickey's heart stopped beating. The unconscious boy gave a gentle sigh as his life ebbed away.

There was no time to think – and only seconds to act. "Cardiac arrest!" yelled Beck, grabbing a scalpel and slicing through the stitches holding the side of the wound. There was just one thing he could do, one terrible option. He pulled aside the bone and muscle and grabbed the boy's heart. It was quivering like a hot bloody jelly.

Straight from the Heart 2

"We've got to try it," said Beck desperately. "If not..."

The porter quickly wheeled Beck's machine, a mass of wires and dials, into the operating theatre and plugged it into the mains.

Beck placed the metal electric paddles to the boy's heart and fired 1,000 volts of electricity. The paddles jumped under Beck's hands but the heart was still, lifeless.

"We're losing him!" shouted the nurse.

Sweat ran down Beck's forehead and into his surgical mask. Once more he was frantically squeezing the slippery heart in his hands. Twenty-five agonising minutes passed, Beck's arms were aching but he dared not stop. More drugs were injected but still the heart would not beat. Perhaps it would be easier to let the boy die, Beck reasoned, knowing he could easily stop. But something drove him on.

"I'll try again," said the surgeon grimly, applying the paddles to the heart with shaking hands. Another jolt of electricity, longer this time, and 1,500 volts made the paddles jump.

There was a long tense silence.

"It's working!" said Beck, his voice hoarse with relief. The heart was pulsing and beating blood strongly and normally as if nothing had happened. And the nurses, the anaesthetist, the whole theatre staff broke into wild cheering.

Later that day Mickey was sitting up in bed.

"I'm starving," he complained. "The food here is shocking."

The nurse smiled, her eyes glistening with happiness and relief. "Well, Mickey," she said, "I think I can safely say we've all had a shocking time."

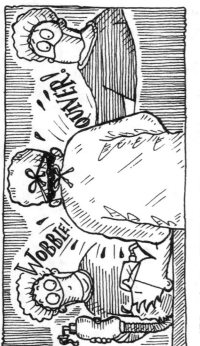

"Ventricular fibrillation!" he snapped. Already he was gently squeezing the heart in his hands – willing it to start pumping blood on its own. Willing the boy to come back to life. For 35 minutes the surgeon frantically massaged the heart between shots of drugs designed to stimulate the muscle – but he knew that he was only buying time. There was just one hope.

"Fetch my defibrillator!" he ordered. "I'm going to try to shock the heart."

He glanced at the white, strained face of the anaesthetist. She was shaking her head.

"But," she protested. "It's never been tested on humans – only dogs."

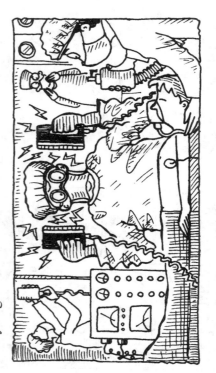

NAME _____ DATE _____

Straight from the Heart 3

● This is your chance to adapt the story into a play.

● You will need to include stage directions (written in brackets) for your actors to follow. It has been started for you...

> **Scene 1:** At the hospital, by the bed of a boy called Mickey.
>
> **Claude Beck:** Now what seems to be the matter, young Mickey?
>
> **Mickey:** *(Looking tired)* Well Dr Beck, I just can't breathe properly and it's really BORING stuck here in hospital while my mates are playing football outside!
>
> **Claude Beck:** Very interesting. *(Turning to his medical students)* You see, Mickey's chest hasn't grown quick enough. He's 14 now so no wonder he can't breathe normally. Am I going to fast?
>
> **Student 1:** No sir!
>
> **Student 2:** No sir!
>
> **Student 3:** *(Dropping his notepad)* Er...well...erm...No sir!
>
> **Claude Beck:** *(Giving Student 3 an angry look, then turning to Mickey)* I wish I could say that the operation will be a cinch, son, but it's an involved procedure...

● Can you finish writing this scene? Use the story to help you.

● Next, see if you can adapt the rest of the story into a play. Split the action into these different scenes.

Scene 2: In the operating theatre

Scene 3: Using the defibrillator

Scene 4: After the operation

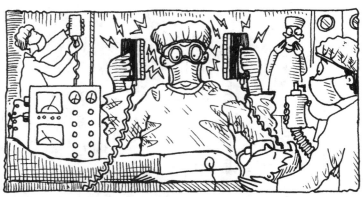

● Top tips: Use your super science skills to design a prop that can act as the defibrillator in your play. Make it light up and buzz to convince your audience!

NAME _____   DATE _____

Straight from the Heart 4

- Acting out your play in front of the class or for an assembly?

- All good plays have a poster – but only very good plays have a poster that lights up!

- Decide what information
  needs to be included...

Title:

Performed by:

Written by:

Date and time:

Location:

My first draft poster design

- Now design a circuit that will light up your poster to tell the audience that something special is happening.

My circuit design

You will need to include:

Bulbs
Battery cells
Switch
Conductors

NAME _____   DATE _____

Lightning Addition Quiz

● Write your answers here:

Answer 1

Answer 2

Answer 3

Answer 4

This quiz is really easy. In fact, you can probably go through it like greased lightning, ha ha. All you have to do is to add up the numbers.

1) How many times does lightning strike somewhere in the world in a second? The answer = $14 + 86$

2) What is the record number of times a person has been struck by lightning? The answer = answer **1)** $- 93$

YES! I'VE BEATEN THE WORLD RECORD!

3) Lightning is hotter than the surface of the sun. By how many times? The answer = answer **2)** $- 1.5$

4) What is the largest number of people ever struck by a single bolt of lightning? The answer = answer **3)** $+ 11.5$

NOT MANY VOLUNTEERS...

...PERHAPS IT'S THE POOR WEATHER

WORLD RECORD ATTEMPT TODAY!

HUH ~ ME LUCKY? ME BOILERSUIT'S RUINED, ME EARRING'S MELTED, AND THERE'S AN OLE IN ME 'AT. AND I BET ME SIDEBURNS WON'T GROW BACK NEITHER. TELL THE PROF I WANT DAMAGES AND DANGER MONEY!

SHOCKING FACT FILE

● Take a look at these fascinating facts...

1 You can make electricity from farts. It's true – by burning methane gas (found in some farts) you make heat which can be used to power generators and make electricity. Methane is also found in rotting rubbish and in the United States there are 100 power stations based at rubbish tips that burn the gas.

2 Lightning is a giant electrical spark. One place that's safe from a lightning strike is inside a metal object like a car. The lightning runs through the metal but not through the air inside – so if you avoid touching the metal yourself you're safe. Much safer than sheltering in an outdoor toilet, for example.

SO YOU CAN MAKE ELECTRICITY FROM FARTS, EH?

3 Sometimes electrical power can surge when the power station pumps out too much electricity. (Imagine a huge wave of power surging into your sockets.) In 1990 people in the English village of Piddlehinton (yes, that's the name) were shocked when a power surge blew up their cookers and TVs.

WOW! THAT'S SO REALISTIC – AS IF OUR TV IS EXPLODING.

IT IS, DEAR.

4 The biggest power cut in history hit the north-east United States and Ontario, Canada in 1965. Thirty million people were plunged into darkness, but luckily only two were killed in the confusion.

● All these statements are true.

● Research some of your own, using books and the Internet.

● Add a few false statements and see if your friends can spot them.

NAME _____ DATE _____

Shocking Inventions 1

● Take a look at this!

● What do you think of the invention?

● List the good points and the dangers.

_____

_____

_____

_____

_____

_____

Bet you never knew!
In Victorian times some people carried lightning conductors on the end of their umbrellas. The device consisted of a metal rod on the spike of the umbrella, with a metal wire attached down which the lightning would run (and hopefully away from the petrified person holding the brolly). It worked in the same way as a full-sized conductor and was designed to keep its owner safe in a storm. But was this a smart idea? I mean, these umbrellas attracted lightning – would you put up with one?

WARNING!
PURCHASERS
WERE ADVISED
NOT TO USE THE
METAL WIRE TO
WALK THE DOG

● How else could you stay safe during a lightning storm?

● Find out by researching books and the Internet.

● Write a safety guide below.

NAME _____   DATE _____

Shocking Inventions 2

- Take a look at this advert.

- What are the advantages of an electric car?

- What are the disadvantages?

- How many other electric inventions can you find?

- Draw and write about them.

- Say whether you think they are successful or not and why.

- Here are some suggestions:

Digital watch

Electric toothbrush

Doorbell

- How many did you find?

NAME _____   DATE _____

SHOCKING ELECTRICITY

My cover